6th Grade Math
Volume 3

© 2013 OnBoard Academics, Inc
Newburyport, MA 01950
800-596-3175
www.onboardacademics.com

ISBN: 978-1494857301

Table of Contents

Compare & Order Decimals

Key Vocabulary

greater than >

less than <

tenths

hundredths

thousandths

Use the symbols to compare these decimals.

 Less than

 Greater than

 Equal to

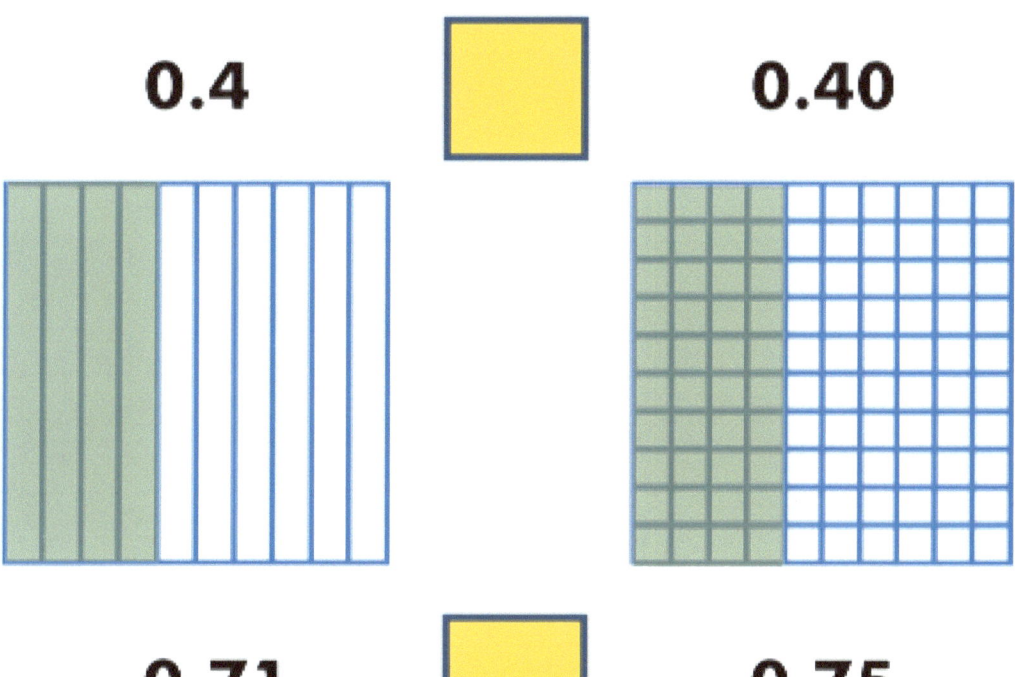

0.4   0.40

0.71   0.75

Write the decimal and then use the symbol to compare them.

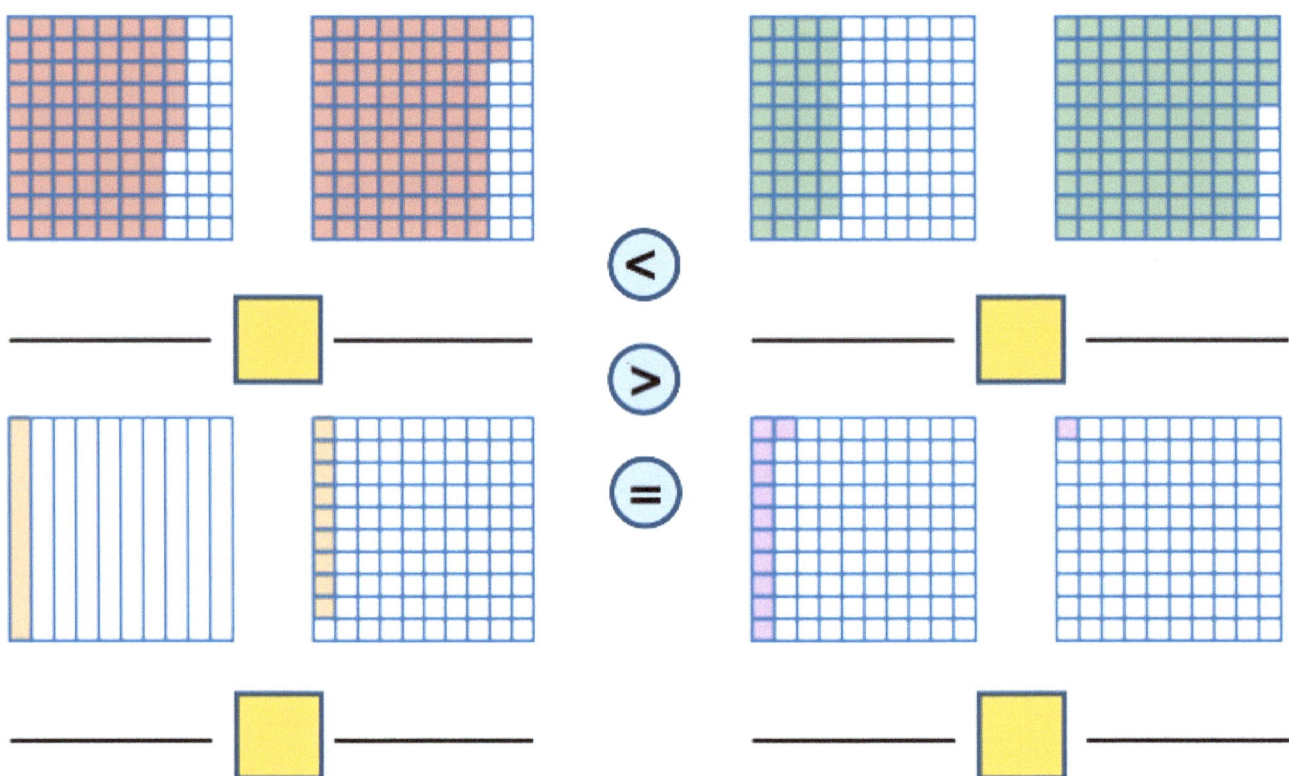

Compare the height of Nancy and Karen.
Write in the correct symbol.
Use the symbols to compare these decimals.

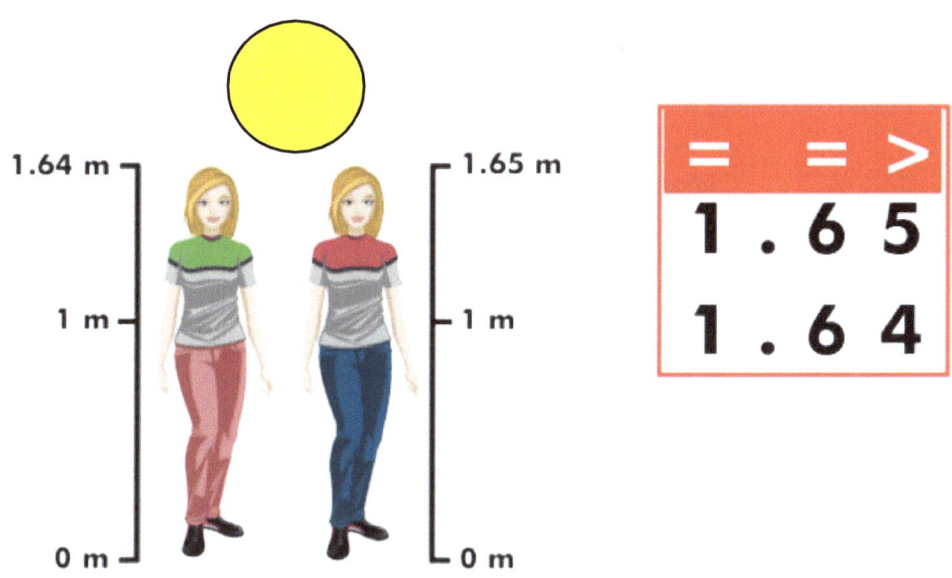

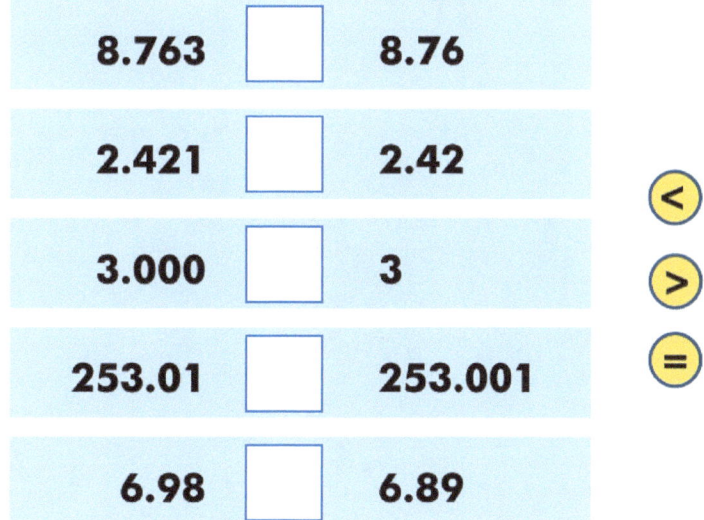

Order these decimals on the number line.

If you have colors, draw the point on the number line. If you don't have colors, draw a line from the decimal to the proper point on the number line.

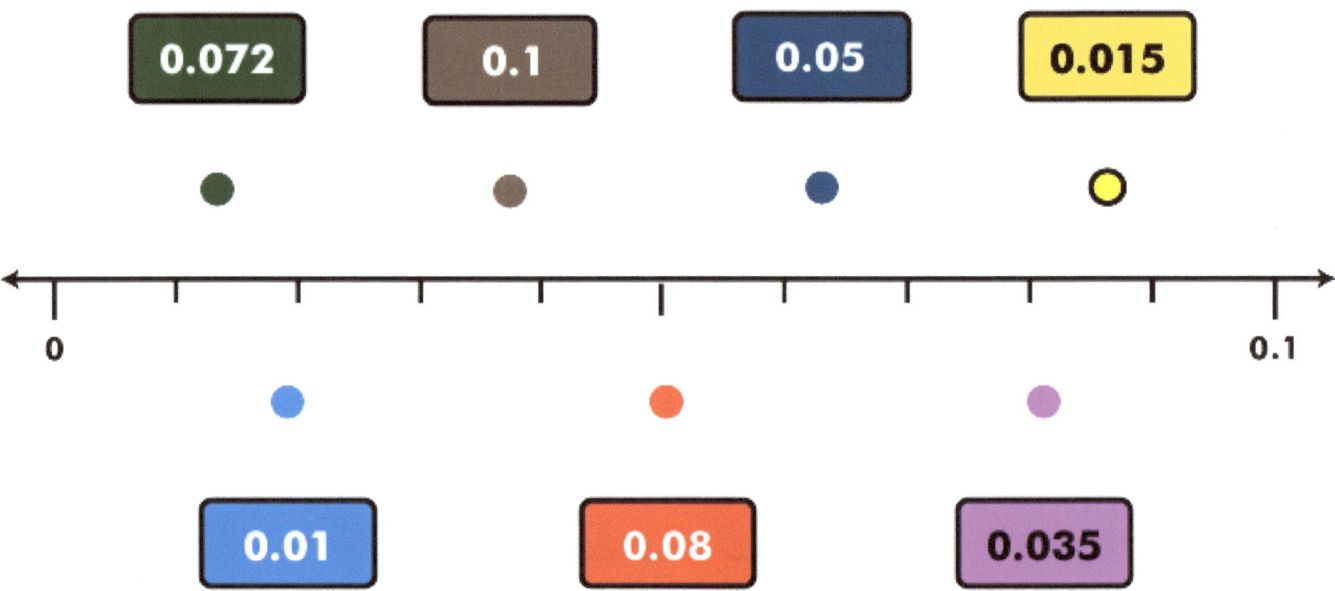

Who won?

Write in the winning times for gold, silver and bronze.

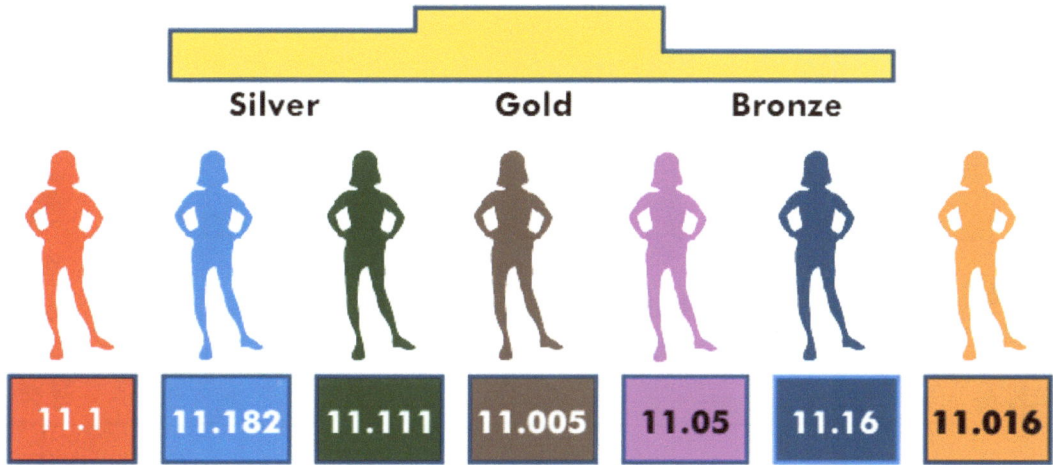

Name_____

Compare & Order Decimals Quiz

1   True or false: 0.4008 > 0.401

2   Which is the correct order (least to greatest):

A	0.007	0.070	0.068	0.09	0.1
B	0.1	0.068	0.007	0.070	0.09
C	0.1	0.09	0.007	0.070	0.068
D	0.007	0.068	0.070	0.09	0.1

3   What is the value for B?

4   What is the value for C?

Adding & Subtracting Decimals

Adding Decimals
Study the illustration below to discover how to add decimals.

$$6.43 + 7.15 + 5.5$$

①	**Line up the decimal points**	6.43 7.15 + 5.50 ————

②	**Add as with whole numbers**	1 6.43 7.15 + 5.50 19 08

③	**Insert the decimal point**	1 6.43 7.15 + 5.50 19.08

Trials for the 4 X 100 m Relay

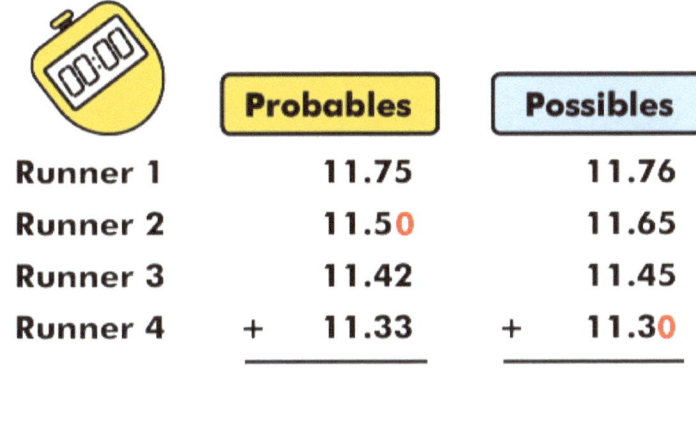

	Probables	Possibles
Runner 1	11.75	11.76
Runner 2	11.50	11.65
Runner 3	11.42	11.45
Runner 4	+ 11.33	+ 11.30

Which team wins the relay, the Probables or the Possibles?

_____

Practice adding decimals.

2.54 + 4.2 + 2.008

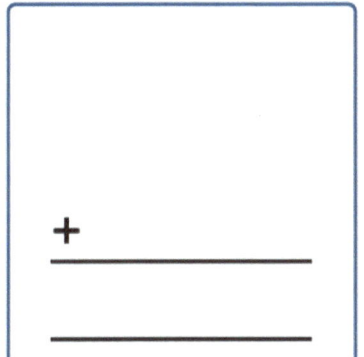

3.26 + 9 + 7.1 + 4.87

Study the illustration below to discover how to subtract decimals.

(1) Line up the decimal points

$$
\begin{array}{r}
286.53 \\
-\ 75.48 \\
\hline
\end{array}
$$

(2) Subtract as with whole numbers

$$
\begin{array}{r}
{}^{4\,1\,3} \\
286.\cancel{53} \\
-\ 75.48 \\
\hline
211\,05 \\
\end{array}
$$

(3) Insert the decimal point

$$
\begin{array}{r}
{}^{4\,1\,3} \\
286.\cancel{53} \\
-\ 75.48 \\
\hline
211.05 \\
\end{array}
$$

(4) Check answer by adding

$$
\begin{array}{r}
{}^{1} \\
211.05 \\
+\ 75.48 \\
\hline
286.53 \\
\end{array}
$$

Subtract the decimals and then check your answer.
Check you answer by adding the answer to the amount subtracted. Does it add up to the original number?

7.54 – 6.3

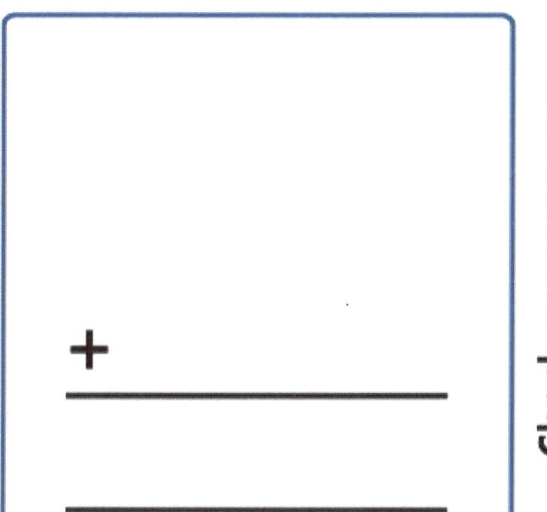

Check your answer

18.341 – 17.85

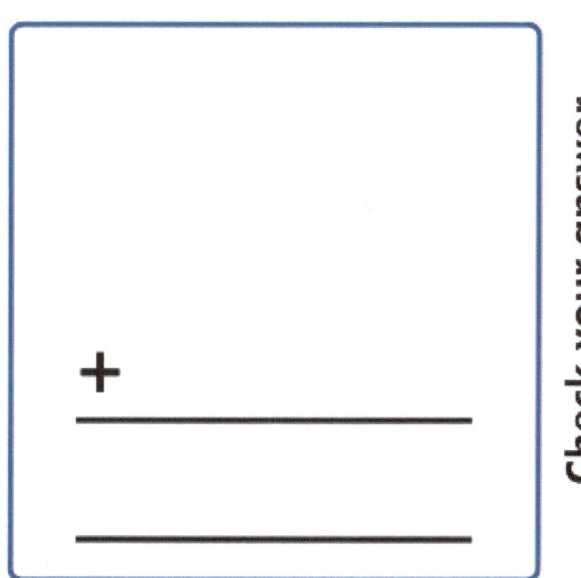

Check your answer

Practice

Coupons for Mrs. Ganguly

Ashima helps out her Mom by clipping money-off coupons. This week at the store, her Mom's bill would have been $286.53, but Ashima has coupons worth $75.68.

What is Mrs. Ganguly's final grocery bill?

Stretch your knowledge.

Grade Amy's homework. How many of her answers are correct?

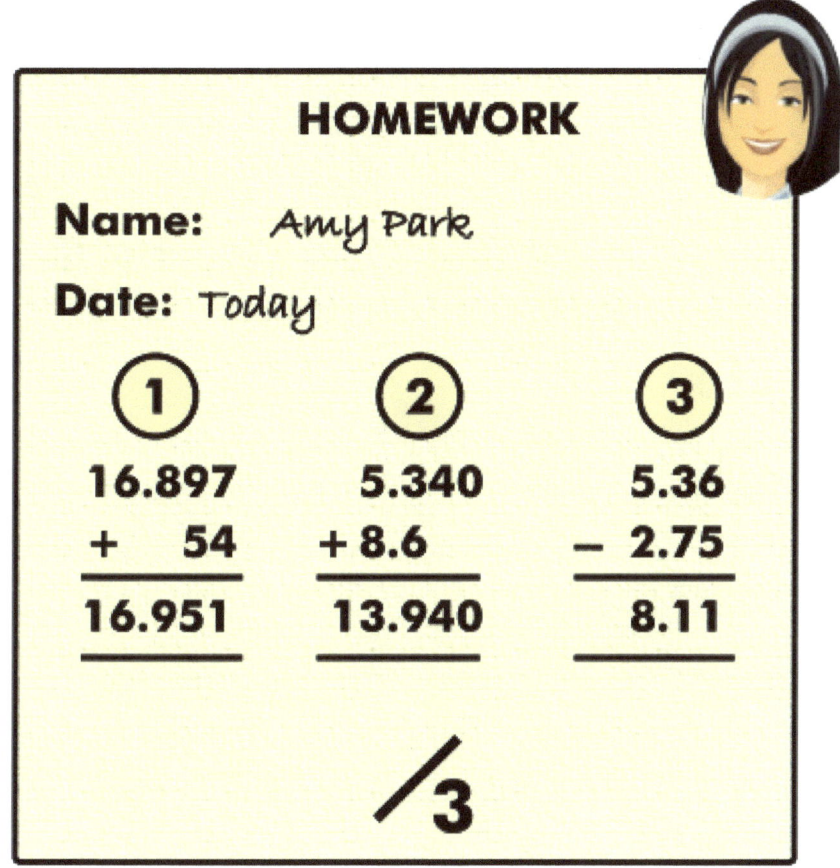

Name_____

Adding & Subtracting Decimals Quiz

1 True or false: 3.5 + 2.02 = 5.7?

2 Alicia purchased three candy bars. The bars cost 87¢, 45¢ and $1.28. How much did she pay?

A $2.60

B $26

C 133.28¢

D $2.48

3 José has $25.45 and spends $9.27. How much does he have left?

4 Here are the 400 m freestyle individual relay times:

46.06   47.05   48   47.21   (times in sec)

What is the team's total time in seconds?

Multiplying & Dividing Decimals

A Multiplication Problem

 Michael has a part-time job at the local coffee shop where he earns $6.72 an hour. On Saturday, he works a 7-hour shift. How much does he earn for this shift?

① Estimate the answer.

☐ x ☐ = ☐

② Multiply as with whole numbers.

```
  6.72
x    7
_____
```

③ Place the decimal point.

```
  6.72
x    7
_____
  4704
```

Hint: Reference your estimate for a clue to placing the decimal point.

Practice multiplying decimals.

 On Sunday, Michael works a $3\frac{1}{2}$ hour shift. His hourly rate on Sundays is $9.78. How much does he earn for this shift?

$$\begin{array}{r} 9.78 \\ \times \quad 3.5 \\ \hline \end{array}$$

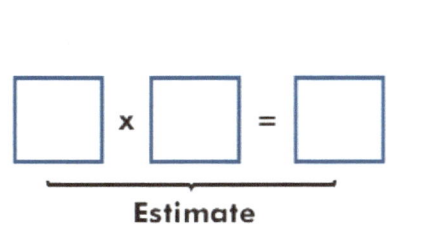

□ x □ = □

Estimate

The number of decimal places in a product, is equivalent to the total number of decimal places in the factors.

_____

Practice placing the decimal point.

1
$$\begin{array}{r} 35.2 \\ \times \quad 16.1 \\ \hline 56672 \end{array}$$

2
$$\begin{array}{r} 90.3 \\ \times \quad 85 \\ \hline 76755 \end{array}$$

3
$$\begin{array}{r} 6.25 \\ \times \quad 1.23 \\ \hline 76875 \end{array}$$

4
$$\begin{array}{r} 1.987 \\ \times \quad 3 \\ \hline 5961 \end{array}$$

Dividing Decimals
Study the process below for dividing decimals.

(1) **Make an estimate.**

(2) **Write the decimal point (dp) in the quotient directly above the dp in the dividend.**

$$4 \overline{)37.36}$$

(3) **Divide as with whole numbers.**

$$
\begin{array}{r}
9.34 \\
4 \overline{)37.36} \\
-36 \\
\hline
13 \\
-12 \\
\hline
16 \\
-16 \\
\hline
0
\end{array}
$$

(4) **Check answer by multiplying.**

$$
\begin{array}{r}
9.34 \\
\times \quad 4 \\
\hline
37.36
\end{array}
$$

Dividing Decimals by Decimals

Look for a pattern in order to find the missing numbers.

Multiplying	Dividing
$3.25 \times 200 = 650$	$650 \div 200 = 3.25$
$3.25 \times 20 = 65$	$65 \div 20 = 3.25$
$3.25 \times 2 = 6.5$	$6.5 \div 2 = 3.25$
$3.25 \times 0.2 = \bigcirc$	$0.65 \div \bigcirc = 3.25$
$3.25 \times \bigcirc = \bigcirc$	$\bigcirc \div \bigcirc = 3.25$

Dividing a Decimal by Decimal

Which division is easier: $65 \div 20 = 3.25$ or $0.65 \div 0.2 = 3.25$?

When dividing by a decimal:

(1) **Change the divisor to a whole number.**

(2) **Multiply the divisor and dividend by a power of 10.**

_____

Rewrite and solve these problems.

(1) **$32.8 \div 0.2$**

(2) **$0.065 \div 0.04$**

(3) **$0.004 \div 2.5$**

Help four friends share this 7 foot sub equally.

TRY OUR NEW
7-FOOT SUB!
only **$7.77**

How long will each piece be?

Estimate

Name_____
Multiplying & Dividing Decimals Quiz

1 True or false? $0.065 \div 0.045 < 1$

2 $3.34 \times 10.8 = ?$

　　A 360.72

　　B 36.072

　　C 33.011

　　D 3.6072

3 If a burger costs $1.69, what's the cost of 4 burgers?

4 $36.8 \div 0.4 = ?$

Fractions, Decimals & Percents

Key Vocabulary

percent

equivalent

Percents, Fractions and Decimals
Try to fill in the blanks.

Total number of digital players:

 **Red digital players:**

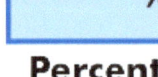      

Number   **Percent**   **Fraction**   **Decimal**

Writing Fractions as Percents

What fraction of the digital players are yellow?

Write this as a percent:

$$\frac{7}{20} = \frac{}{100} = \boxed{\%}$$

per cent
= per 100

Practice Questions

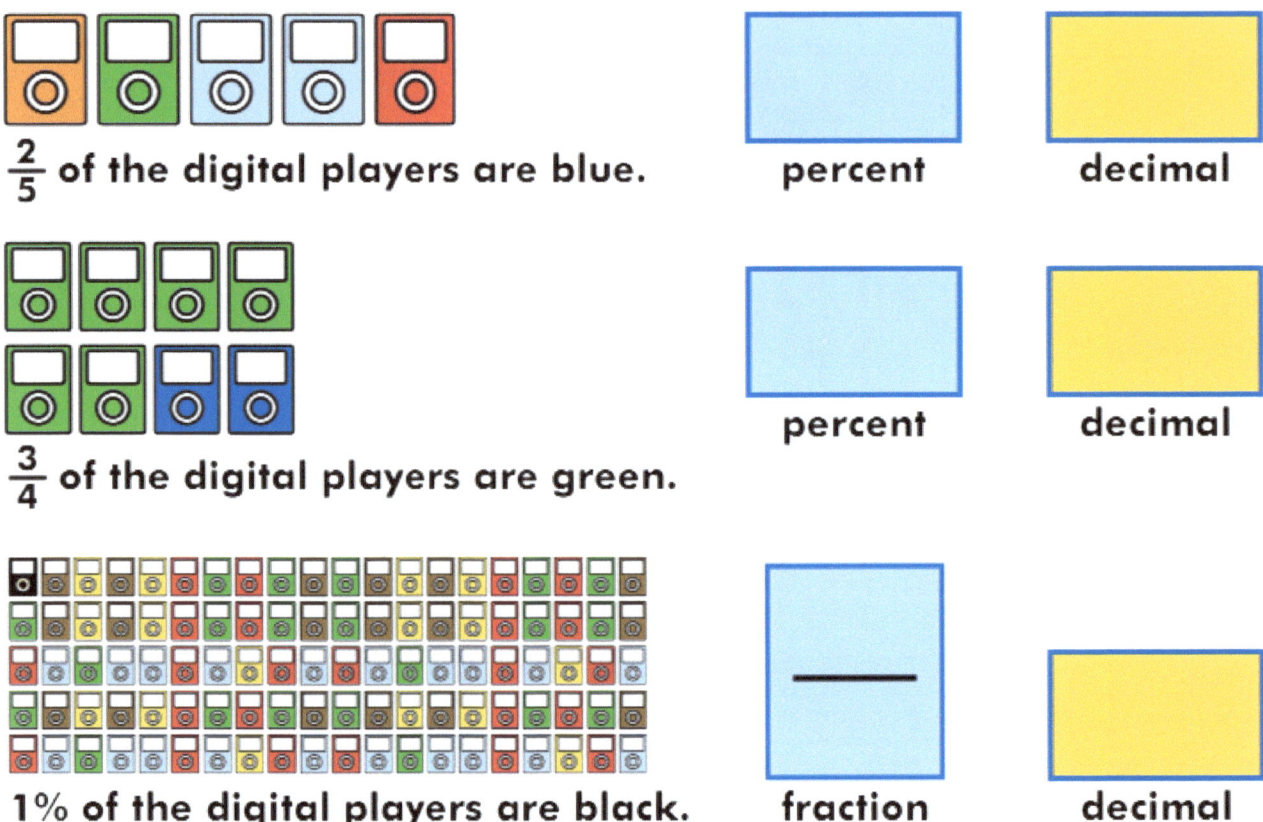

$\frac{2}{5}$ of the digital players are blue.

percent   decimal

$\frac{3}{4}$ of the digital players are green.

percent   decimal

1% of the digital players are black.

fraction   decimal

Find the missing values.

Percent	Decimal	Fraction
	0.35	$\frac{7}{20}$
70%		
99%	0.99	
7%		$\frac{7}{100}$
	0.02	

Name_____

Fractions, Decimals & Percents Quiz

1 True or false: 1% = 0.001?

2 Which of the following statements is *not* true?

   A $\frac{1}{20}$ = 5% = 0.05

   B $\frac{3}{100}$ = 3% = 0.3

   C $\frac{3}{5}$ = 60% = 0.6

   D $\frac{3}{4}$ = 75% = 0.75

3 Write 55% as a decimal.

4 Write 0.07 as a percent.

Newburyport, MA 01950

1-800-596-3175

OnBoard Academics employs teachers to make lessons for teachers! We create and publish a wide range of aligned lessons in math, science and ELA for use on most EdTech devices including whiteboard, tablets, computers and pdfs for printing.

All of our lessons are aligned to the common core, the Next Generation Science Standards and all state standards.

If you like our products please visit our website for information on individual lessons, teachers licenses, building licenses, district licenses and subscriptions.

Thank you for using OnBoard Academic products.